Octo has such gangling arms,
she wants to cover them with charms.

First she gathers spiral shells,
then looks around and finds some bells.

How many charms
on her eight gangling arms?

And here's a trunk with silver bars,
and golden medals shaped like stars.

With all these charms her arms are sore,
so why does Octo look for more?

"Look, I'm starting a trinket store!"
TRINKETS FOR SALE